THE CONDITION
called poetry

THE CONDITION
called poetry

Kadi M. Spear

TATE PUBLISHING
AND ENTERPRISES, LLC

Published by Tate Publishing & Enterprises, LLC
127 E. Trade Center Terrace | Mustang, Oklahoma 73064 USA
1.888.361.9473 | www.tatepublishing.com

Tate Publishing is committed to excellence in the publishing industry. The company reflects the philosophy established by the founders, based on Psalm 68:11,
"The Lord gave the word and great was the company of those who published it."

Book design copyright © 2014 by Tate Publishing, LLC. All rights reserved.
Cover design by Gian Philipp Rufin
Interior design by Manolito Bastasa

Published in the United States of America

ISBN: 978-1-63418-137-2
Poetry / General
14.10.04

CONTENTS

Part 1 — Life Is a Song Sung By a Poet

Part 2 – Poems without Meanings Often Hold Deeper Truths

Part 3 – Poetry Is 10 Percent Life and 90 Percent Leisure

LIFE IS A SONG SUNG BY A POET

These poems are poems about my life. I will not say the meaning behind them; that is something that I hope you shall be able to gather for yourself. Make your own meaning or understand mine, whichever works for you. I just hope that you will take a minute to understand them and the story that they portray.

From Sidewalk Drawings

I am from veterinarian dreams,
From jungle gym mornings and hopscotch nights.
I am from sidewalk drawings,
In the chalk the guarded the dreams of a child.
I am of Amelia Earhart books
And the echoing songs of Disney films.
I am in the memories
Falling all around me,
Pieces rained down from a time long gone.
I am in the Polly Pockets,
Junie B. Jones, and Tamagotchies.
Bubble Tape, *Nancy Drew*, and *Rugrats*,
Boxcar Children and Lincoln Logs,
These are the things I am of.
Full House and lemonade stands,
Summer games with summer friends, this I am of.

The smell of grass under the rain of the sprinklers,
Hot summer days and rosy cheeks,
From this and those I am of.
The days of grandparent hugs, this I am of,
The smell of Old Spice and a cigarette.
From tales of a childhood that was not mine,
Passed on from aging minds to little souls.
The place in my mind
Where I keep these memories of mine.
An old scrapbook or two,
A box or a trinket tucked away,
Things to come and things behind,
These, most of all, are the things I am from.

Ask Me

Tell me how it feels
To be so innocent,
A child of five,
Or, rather, I should tell you.
Braving the change,
A movie, the scurry of a teacher, ask me.
Excitement, the wonder of secrecy.
A mother, a tale as different as age,
Her understanding, deep-seated fear.
Her children, so far away,
Unsure of their fate,
Unsure of her own.
A nation so humbled,
And a mother so afraid, ask me,
Or rather ask her.

There Is This Girl You Never See

There is this girl you never see,
Nary a word she speaks.
And, still, she yearns to be noticed,
For once, to be in the spotlight and not shy away.
But this girl, this invisible girl,
She feels these feelings she can't understand.
So the razor she takes to restore,
But with every cut, she digs a deeper void,
A grave to lay herself in.
Those around her, they care for her,
But to this, she is blind.
A world of hatred and pain she lives in
Until her angels come for her,
Ripping her from the grave of her own flesh.
Until maybe, just maybe
She sheds the guilt and hurt,
And once again rejoins the life she was meant to live.
Until then, however, she remains.
The girl you never see.

Cut Deeper Than Blood

One breath,
One second,
One chance
Wasted.
No turning back now,
Your choice is made.
Pick up the razor,
Put down your hope.
Cut deeper than blood,
Last longer than high,
Feel more than nothing,
Feel less than everything.
Keep going,
Don't stop,
Can't stop.
You won' let go,
It won't let go.
Cut deeper than blood,
Last longer than high,
Feel more than nothing,
Feel less than everything.

Rose and Thistle

You were sweet.
I was naïve.
A bleeding heart
In a world of vulnerability.
You came.
A savior to my soul,
A rose to my thistle.
Moments melded to months,
The rose became the thistle,
And the thistle the blood.

Left Me Laying

You came to me
So innocent,
So pure.
A child in need of love,
A lover in need of hope.
You broke me
To nothing,
The dust of a memory.
Left me laying
In a pool of blood and regret.
Stole my innocence
With a lustful mind
And impure hand.
Left me laying
In the bed of my own hatred.

Ruiner

Ruiner—
Of hearts,
Of souls,
Of hope.
Thief—
Of second chances,
Of easy explanations,
Of peaceful dreams.
Beggar—
Of belief,
Of hopeful souls,
Of a common enemy.

Soul of Sighs

Eyes so black
Reflect the night,
Steal a soul.
Keeper of hope,
Chest of light.
Trapper of minds,
Chest of whispers,
Lover of broken hearts.
Eyes so black
Reflect the night,
Steal a soul.

But I Love You, Baby

Lost to faith
And lost to you.
White words
On deaf ears.
"Never again, baby."
The familiar sting.
Words whip harder than hands.
"I promise, baby."
An unwelcome touch,
Harsh words,
And a harsh reality.
"I'm sorry, baby."
Wake up.
Stay asleep.
What does it matter?
What do you matter?
"But I love you, baby."

End. Now.

Friend,
You call yourself.
Honest,
You claim to be.
To the end,
Now.
Traitor,
Your truth shows you to be.
Liar,
Your friendship proved to be.
This is the end,
Now.

Menacing Agenda with Backwards Smiles

I won't hear your lies,
Harsh words from evil souls.
A menacing agenda
With backwards smiles.
Come to me,
All your problems I will solve.
Come to me,
Use me, walk all over me.
Come to me,
With my trust I hand you my life.
Come to me,
Oh, friends of mine.
He smiles,
A snake in the grass.
You believe,
Loyalty falls.
I see now,
You never were
Friends of mine.

Tiny Steal Heart

Cold metal
Against hot fingers.
A past,
A future,
Tied to words,
Tied to hand.
A little ruby,
Tiny steal heart,
Words,
Promise.
Lost.

Fall between Fingertips

The empty room,
A sullen reminder
Of the silent scream.
The stale smell,
So familiar yet so foreign.
Thin pages
Fall between fingertips.
The wooden pew,
Cold against the palm
Of the ever-breaking heart of a believer.
Thin pages.
Silent winds.
Hair blown to tears.
Thin pages
Fall between fingertips.

Bound Tightly Around

Holding my dreams,
My past,
My hopes and future,
Bound tightly around.
A promise.
"Love Waits,"
Lust won't.
You came,
You took.
Promise.
Fall swiftly,
Metal hits the floor,
Ringing in heart and ear.

A Woman from the Girl

There is a girl you never see,
Hiding behind fear and secrecy.
On the world, she has set her eyes,
Never knowing she was guaranteed the prize.
A woman from a girl she becomes.
Her voice no longer trembles,
Her hand no longer shakes
For fear, she will embrace.
Her fight soon will end.
In victory, she claims the prize.
Eternal life in her Lord
And grace she will accept.
No longer hurting, no longer blind
To the miracles and angels around,
To the world in which she lives.
As she walks amongst angels
She learns to become the woman from the girl.

Best Friend

A word meaning much more than it seems,
A dictionary will do no good,
An inquiry will prove what is true.
A secret keeper,
A dreamcatcher,
An angel to your soul.
What luck would find a best friend in the best of times?
What luck could keep, when the water is too deep?
A best friend in me, you will always find,
For I know in you, I will always have mine.

Two Years Since

It's been two years,
I find myself saying.
Two years since recovery.
February seventeenth
Two years.
You're better.
Good for you.
Smile
Because they'll never understand.
Smile
Because the scars are covered.
Smile
Because you have to.

Recovery

The beep of the phone.
One calendar event for today.
February,
It's the seventeenth,
A thought pushed away.
Recovery,
It gleams.
Look what you've done.
Recovery.
Click,
It disappears.
Recovery.
Look what you've done.

POEMS WITHOUT MEANINGS
OFTEN HOLD DEEPER TRUTHS

These poems are some poems that I wrote just for the heck of it. All poems, of course, are written with a meaning; as for these poems, even I may not know their meaning. These are just some poems I wrote because I was in the mood for poetry. Some of them may be true and they may have a meaning. Either way, I hope you enjoy them.

At the End of the Road

It's a new day,
But it's all the same.
Sun gleams off the snow,
Sparkles shine,
A disco ball from the sky.
Cars pass.
Do you wonder like I do?
Are they a family?
At the end of their road,
Do they wish to keep walking?
Slippery slope.
A porch light flickers.
Dogs bark,
Cats meow,
Parents fight,
Brother yells.
Cars pass.
Do you wonder like I do?
Are they a family?
At the end of their road,
Do they wish to keep walking?
Is this a home?
Or just a house?

I Dream

My dreams aren't of beaches,
Palms trees, and margaritas.
I don't dream of fancy cars or diamond rings,
Botox women, or fancy things.
Far-off places flicker through my thoughts.
My dreams?
I dream of a place
Filled with warmth,
A place where frowns are strange
And dreams nurtured.
I dream of a place
Where the black sheep is revered,
Where the bookworm isn't questioned,
And where everyone listens.
I dream of a place called home.
I dream of a place to dream.

People in Colors

I see people in colors.
Not color.
Colors.
Red, green, blue, purple.
Colors.
Souls are sheeted white.
Dreams are shades of blue.
Hurt is shrouded in greens.
And memories shine yellow.
I see people in colors,
Layers,
Like onions,
Each with their own color.
He has been hurt.
She has dreams.
She is blue.
He is green.
Layers.
Colors.
Like onions.

Home

Home—
A loaded word.
A loaded aspiration.
Where is home?
A building.
A house.
A feeling.
A person?
Home.
Does anyone really have a home?

Why Try? Try Harder

Aspiration—
A hard thing to have,
A stomping ground.
Unattainable.
Unimaginable.
Unrealistic.
So they say.
Why try?
Try harder.
Why try?
Try harder.
Try harder.

Walk with Lies, Stand with Drama

Blah, chatter, talk,
They all will.
Whisper, glance, a sly smile.
Disease called
Drama,
Routine called
Life.
Smile,
Fake it.
Love it,
Hate it.
Speak, chatter, talk,
Love the rumors.
Walk with lies.
Stand with drama.
Fight with life.

The Final Sun Sets

The final sun sets, the final wind blows.
Every tomorrow a yesterday.
The days, the months, the years fly by
On a colored wind of tomorrow.
Lingering, waiting, tedious time
Sweeps you up on golden ray of sun.
Carries you softly off to sea,
Drifting slowly, slowly away.

The Moon

The moon,
How innocent to love the moon.
For there is no love as the love of the condemned.
Such power does it hold.
To destine and bring the destined deep into the night.
To leave the flesh behind
And run with the spirits of the moon.
To swim deep into its light
If only to touch the dark.
Flow into the tune of the tides.
Ride the waves of time.
Fall into the trance of the generations
Of former and future.
To be held and to hold
In the stories of the galaxy.
Discover the secrets of eras
As you run to the moon and back.

Liberty, My Lady

Liberty, my lady,
A humbled servant I am to thee.
I walk the lonely streets.
A child to thine unknown,
Singing the song of the anonymous.
Liberty, my lady,
A humbled servant I am to thee.
I run the crowded mile.
A child to thine renowned,
Singing the song of the legendary.
Liberty, my lady,
A humbled servant I am to thee.
I crawl the rock-strew, unsteady streets.
I soar the carefree unwavering mile,
Singing the song of the broken,
forgotten, nameless assemblage,
Singing the song of the unbroken,
infamous, memorable congregation,
Singing the song of America.

An Ode to Authors

It is more than the paper.
It is more than the pen.
It cannot be faked.
The talent it takes.
Many will try,
Few will succeed.
But let us not dwell
On the struggles we face.
Where would we be
Without those who tried?
Such a dull place
Would the world be.
So here is to the people
Who took the chance in stride.
So let it be known.
Here is to the writers.

A Forgotten Child

Darkness sets in.
Once again he is alone.
She left him there,
The corner of a stuffy room,
Where no one can hear him cry.
Crying, alone, confused.
In the dead of night,
He sits alone knowing this just is not right.
He cannot understand
Why she will not hear,
Why her heart just cannot care.
Though he never asks why,
He never flees,
Waiting until he knows
He will spend another night alone.

The Ballad of a Teen

Her heart is slowly breaking
With every word he isn't saying,
Every smile that isn't hers,
Her spirit's gently shaking
As she struggles to go on.
And with every smile she's faking,
Every laugh she doesn't mean,
Her life is slowly crumbling as she falls to her knees.
The tears are pouring down.
Oh, but she doesn't care.
No, she doesn't care.
She's falling.
Oh, she's falling.
She knows it isn't right.
She knows she shouldn't care.
'Cause for him there was nothing there.
But she's falling…down.
The day will come,
And she'll see
That you've got to crawl to learn to walk.
You've got to fall to learn to stand.
You can't run; you can't fly
Without the hurt, without the pain.
In the end, it's all the same.

Ash Meadows

She's wild, and she's free,
But she hides the fear underneath.
She doesn't know that she's wonderful.
She's strong, confident,
Lovely, funny,
And unique.
Everything she needs to be.
But secretly, she wonders.
Secretly she thinks,
Maybe, just maybe
She's wrong.
But she's strong, confident,
Lovely, funny,
And unique.
Everything she needs to be.
Even if she doesn't fully see.

The Royal Hill

She's young, but she doesn't know
She has so much time to grow.
She worries about what they think,
Finding every flaw.
But she doesn't know
She's perfect.
She just has to let it show.
Hiding behind flaws, too scared to see
That the faults she sees just aren't real.
She's pretty, she's young.
She's smart, she's talented.
But she doesn't know.
She's too scared to grow,
To shine,
To be everything she could be
And get everything she deserves.
She worries and she hurts,
Scared to fight for what she deserves.
She'll never know
That she's perfect.

POETRY IS 10 PERCENT LIFE AND 90 PERCENT LEISURE

These poems were inspired by novels, short stories, or novellas that I am currently writing or have finished writing or had the idea for. The inspiration may not be clear because you probably do not know the novel, short story, or novella they were inspired by. Nonetheless, I hope you enjoy them.

Destiny

Destiny—
A simple word,
A meaning that does no justice
To the power it wields,
The grasp it holds on our soul.
Futile is it to fight such an authority as this
If a tale of woe is not sought.
Life will show her the ultimate choice
Never truly was.

Everance

Above all else, I am going to succeed,
even at the cost of others.
Those who get in my way will never prosper.
The liar, a fraud, so lost was the girl behind the mask.
So forgotten, so different,
Unsure of who she was,
So she became who she thought she should be.
She became falsely foolish,
Bubbly and vibrant,
Unafraid of social limitations.
Some called her fake; some thrived off her power.
As her power over her peers grew,
So did her self-confidence.
Slowly, the girl underneath faded,
And the arrogant façade prospered.
Never was she without success,
Never was she without her façade.
But what did it cost
To never stop and ask,
To never stop and care,
To never stop.
At what cost?

The new me doesn't need to bring
others down to be happy.
I have thrived at the bottom, and I have faltered at the top.
Now I know the true path to success.
An altercation with reality.
The girl within fighting to be heard.
And the arrogance receded.
The happiness she longed for was discovered.
The intelligence she hid,
The kindness the feared,
And the true popularity she longed for
Found her long before she knew she was looking.

Success still found her,
And no longer did she fear
Her reasonings or the reasonings of life
But embraced them and the existence they brought her.

Mother: A Tale of Tara

A woman from whom the depths of love are boundless.
A woman from whom the solutions to our
tragedies lie in the simplest of words or gesture,
Which we pass off never to understand
their full meanings.
A woman whose personal tragedies are forever
buried so far beneath the surface we shall
never comprehend their full weight.
A woman to whom we owe everything, to
whom the world owes everything
Yet asks for nothing in return.
She takes in the hopeless refugees,
The angry soldiers returning from the simplest of wars.
Heals the tender wounds of the beaten fighter,
Helps you soar high with the heroes.
In the end, when the boy matures to the man,
The girl to the woman,
She will always protect you from
the hurt inevitably to come
And make everything better long,
long after you think she can.

Revolutionize

Shall I look about me to the world?
When all there is to see
Is a world dying about me.
No morality, no shame, no care,
The people they shout, they cry out in vain,
"To fix it myself just would not be fair,"
So as content to grumble as we all be,
Living in a world of should be.
The changes sought out,
The picture repeated.
Wars waged, poverty increased, problems ignored.
Are we not ready to wage our own war?
To fight for the change,
To be that which others critique.
It shan't be easy, nor quick, nor fun.
The soldier, the struggle, the fight,
With even one person, the goal is in sight,
But one cannot wait until it is done."
As has been said,
As will be said,
Don't wait for change...be change.

Spirit

So young, so sweet,
The young girl just learning to love.
He who haunted her dreams and plagued her childhood,
Fueled her desire and ran with her wishes.
So now, secretly she hunts, but still the years pass.
The children she saves,
The people she reunites
Ease her conscience not.
The ignorant people of the town
Call her a hero.
In her mind, she's no hero.
Her false façade, bravery, tenacity
Only aid in her fury
And add to the secrecy of her fear.
The family who forgave
And the daughter who couldn't forget.
They call her spirit.
How far can she go before she loses hers?

The Grasp of Power

The secret so simple in appearance,
The knowledge so deep,
The greatest of minds have not the crevasse to encompass.
The youth, the power, the destiny.
So different in their weight
Yet so similar.
Youth: a time for infinite adventure.
So innocent, so harmless in nature.
Reality is all too quick to step up.
Power, as it is to corrupt,
It is made to build.
The pure, the innocent, the righteous,
Those young enough to grasp it,
Those old enough to know
The destiny it sends forth.

The Miracle of Rayne

Most people cannot see
The suffering he hides
Deep underneath.
The surface of gloom and hate,
The tormented spirit
Of a once-vibrant boy.
He loved and laughed.
They came and went,
Leaving him to fend and to feed.
The burden they saddled him with
Threatened to drown him.
But those little miracles
Kept him afloat.
Long years came,
And long years went.
The weight grew stronger,
And he grew weaker.
The little miracles no longer saved him

But rather innocently brought him down.
His solution was pain and isolation
Until the day he laughed at love,
And it slapped his face
With a beautiful girl.
Understanding and kind,
She did not prod,
Just simply smiled and silently understood.
She could tell he was hurting.
No matter how he hid that he had fallen,
She lifted him back up.
No one could ever take that from him,
The memories of a wonderful girl life brought him.
With music playing softly in the back
And the sun beating down,
With waves sloshing in the distance—
That was the summer
He fell in love
With the miracle of Rayne.

The Trees of Truth

Blue sapphires set upon light sheets of soul.
A river of black waves rest upon a country of essence,
Light as a feather stretching 'cross the land.
The casting shadow of spirit,
The fire burning deep beneath,
The rolling hills of truths,
The pearls so white,
E'er enclosed in boundless jungles of exploration.
A tragedy so great, so ceaseless, so continuous,
Shattering the sapphires; darkening the soul sheets.
Drain the river, conquer the shadows, douse the fire,

The eternal burn of exploration scuffed
As the pearls
engrave the trees of the truth,
Forever in the heart of the silently broken.